U.S. Department of Transportation
National Highway Traffic Safety Administration

DOT HS 811 697

January 2013

Occupant Restraint Use in 2011: Results From the National Occupant Protection Use Survey Controlled Intersection Study

DISCLAIMER

This publication is distributed by the U.S. Department of Transportation, National Highway Traffic Safety Administration, in the interest of information exchange. The opinions, findings, and conclusions expressed in this publication are those of the authors and not necessarily those of the Department of Transportation or the National Highway Traffic Safety Administration. The United States Government assumes no liability for its contents or use thereof. If trade names, manufacturers' names, or specific products are mentioned, it is because they are considered essential to the object of the publication and should not be construed as an endorsement. The United States Government does not endorse products or manufacturers.

Suggest APA Format Citation:

Pickrell, T. M., & Ye, T. J. (2013, January). *Occupant Restraint Use in 2011: Results from the National Occupant Protection Use Survey Controlled Intersection Study*. (Report No. DOT HS 811 697). Washington, DC: National Highway Traffic Safety Administration.

Technical Report Documentation Page

1. Report No. DOT HS 811 697	2. Government Accession No.	3. Recipient's Catalog No.
4. Title and Subtitle Occupant Restraint Use in 2011: Results From the National Occupant Protection Use Survey Controlled Intersection Study		5. Report Date January 2013
		6. Performing Organization Code NVS-421
7. Author(s) Timothy M. Pickrell[*] and Tony Jianqiang Ye[†]		8. Performing Organization Report No.
9. Performing Organization Name Mathematical Analysis Division, National Center for Statistics and Analysis National Highway Traffic Safety Administration U.S. Department of Transportation 1200 New Jersey Avenue SE. Washington, DC 20590		10. Work Unit No. (TRAIS)
		11. Contract or Grant No.
12. Sponsoring Agency Name and Address Mathematical Analysis Division, National Center for Statistics and Analysis National Highway Traffic Safety Administration U.S. Department of Transportation 1200 New Jersey Avenue SE. Washington, DC 20590		13. Type of Report and Period Covered NHTSA Technical Report
		14. Sponsoring Agency Code
15. Supplementary Notes [*] Mathematical Statistician, Mathematical Analysis Division, National Center for Statistics and Analysis, NHTSA [†] Statistician, Bowhead Systems Management Inc., contractor working at NHTSA		

Abstract

This report presents results from the 2011 National Occupant Protection Use Survey (NOPUS) Controlled Intersection Study. NOPUS is the only nationwide probability-based occupant restraint use survey. The National Center for Statistics and Analysis of the National Highway Traffic Safety Administration conducts this survey annually. The 2011 NOPUS found that restraint use for all children from birth to 7 years old increased significantly from 89 percent in 2010 to 91 percent in 2011. Significant increases in child restraint use in 2011 occurred among children traveling in vans and SUVs, in the Northeast, and during weekends. Among occupants 8 and older, seat belt use in front seats continued to be lower among 16- to 24-year-olds than other age groups. Seat belt use in rear seats stood at 74 percent in 2011.

17. Key Words Seat belt use, child restraint use, observational survey, demographics, rear-seat seat belt use, occupant protection, NOPUS, controlled intersections, car seat		18. Distribution Statement Document is available to the public from the National Technical Information Service www.ntis.gov	
19. Security Classif. (of this report) Unclassified	20. Security Classif. (of this page) Unclassified	21. No. of Pages 29	22. Price

Form DOT F 1700.7 (8-72)   Reproduction of completed page authorized

TABLE OF CONTENTS

Executive Summary v

1. Introduction 1

2. Demographic Results 2
   Age 2
   Gender 3
   Race 3
   Presence of Passengers and Seat Belt Use 4

3. Seat Belt Use in Rear Seats 6
   Seat Belt Use in Rear Seats Versus in Front Seats 6
   State Laws and Rear-Seat Belt Use 6

4. Child Restraint Use 9
   Child Restraint Use Among All Children Age Under 8 9
   Child Rear Seat Placement 9
   Child Restraint Use by Region 10
   Child Restraint Use by Time of Week 11
   Child Restraint Use by Vehicle Type 11
   Child Restraint Use by Driver Belt Status 12

5. NOPUS Methodology 16
   Sample Design 16
   Data Collection 17
   Estimation 18

6. References 19

   Appendix: Definitions 20

LIST OF FIGURES

Figure 1: Seat Belt Use by Age for Occupants 8 and Older in 2010 and 2011 2

Figure 2: Seat Belt Use by Age for Occupants 8 and Older, 2002-2011 3

Figure 3: Seat Belt Use by Gender for Occupants 8 and Older, 2002-2011 3

Figure 4: Seat Belt Use by Race for Occupants 8 and Older, 2005-2011 4

Figure 5: Passenger Effect on Seat Belt Use for Occupants 8 and Older, 2004-2011 4

Figure 6: Seat Belt Use by Seating Position for Occupants 8 and Older, 2004-2011 6

Figure 7: Seat Belt Use in Rear Seats by State Law Type for Occupants 8 and Older, 2005-2011 7

Figure 8: Child Restraint Use Among Children Under Age 8, 2002-2011 9

Figure 9: Child Rear Seat Placement, 2002 - 2011 10

Figure 10: Child Restraint Use by Region in 2010 and 2011 10

Figure 11: Child Restraint Use by Region, 2004-2011 11

Figure 12: Child Restraint Use by Time of Week in 2010 and 2011 11

Figure 13: Child Restraint Use by Vehicle Type in 2010 and 2011 12

Figure 14: Child Restraint Use by Driver Belt Status, 2004-2011 12

LIST OF TABLES

Table 1: Passenger Vehicle Occupant Seat Belt Use by Demographic and Other Characteristics 5

Table 2: States With Laws Requiring Seat Belts Be Used in All Seating Positions 7

Table 3: Seat Belt Use in the Rear Seat of Passenger Vehicles, by Major Characteristics 8

Table 4: States With Laws Requiring Children 5 and Younger Be in the Rear Seat 10

Table 5: Child Restraint Use in Passenger Motor Vehicles, by Major Characteristics 13

Table 6: The Percent of Children Who Rode in the Rear Seat, by Major Characteristics 14

Table 7: Child Restraint Use in Passenger Motor Vehicles, by Age and Other Characteristics 15

Table 8: Sites, Vehicles and Occupants in the 2011 NOPUS 17

Executive Summary

The National Occupant Protection Use Survey (NOPUS) is the only nationwide probability-based survey of seat belt use (for occupants 8 and older in both front and rear seats), motorcycle helmet use, child restraint use (for children less than 8 years old), and driver electronic device use in the United States. The National Center for Statistics and Analysis of the National Highway Traffic Safety Administration conducts this survey annually. Two sub-surveys: the Moving Traffic Survey and the Controlled Intersection (CI) Study comprise the NOPUS.

In the CI study, occupants of passenger vehicles without commercial or government markings are observed from the roadside at intersections controlled by stop signs or stop lights. Only stopped vehicles are observed to allow ample time to collect a variety of information required by the survey. NOPUS derives its estimates of seat belt use in rear seats, child restraint use, driver electronic device use, and demographic characteristics of vehicle occupants from the CI study.

This report presents results of occupant restraint use from the 2011 National Occupant Protection Use Survey Controlled Intersection Study. NHTSA will publish driver electronic device use results in a separate report.

The following are some of the major findings from the 2011 NOPUS Controlled Intersection Study:

Child Restraint Use (For Children from Birth to 7 Years Old):
- Restraint use for children from birth to 7 years old increased significantly to 91 percent in 2011 from 89 percent in 2010.
- Restraint use for children in vans and SUVs increased significantly to 97 percent in 2011 from 92 percent in 2010.
- Restraint use for children in the Northeast increased significantly to 94 percent in 2011 from 89 percent in 2010.
- Restraint use for children traveling during weekends increased significantly to 94 percent in 2011 from 87 percent in 2010.
- The rear seat placement rate for children age 4 to 7 increased significantly to 92 percent in 2011 from 89 percent in 2010.

Front Seats Belt Use (Among Occupants 8 and Older):
- Seat belt use continued to be lower among 16- to 24-year-olds than other age groups.
- Seat belt use continued to be lower among males than females.
- Seat belt use continued to be lower among black occupants than occupants of the other race groups.
- Seat belt use continued to be lower among drivers driving alone than among drivers with passengers.

Rear Seats Belt Use (Among Occupants 8 and Older):
- Seat belt use in rear seats in 2011 stood at 74 percent.
- Seat belt use in rear seats in 2011 continued to be higher in the States with laws requiring belt use in all seating positions (83%) than in the States requiring belt use only in the front seat (67%).

1. Introduction

The National Occupant Protection Use Survey is the only nationwide probability-based survey of seat belt use (for occupants 8 and older in both front and rear seats), motorcycle helmet use, child restraint use (for children less than 8 years old), and driver electronic device use in the United States. The National Center for Statistics and Analysis of the National Highway Traffic Safety Administration conducts this survey annually. Two sub-surveys: the Moving Traffic Survey and the Controlled Intersection (CI) Study comprise the NOPUS.

In the MT survey, front-seat occupant shoulder belt use data and motorcyclist helmet use data are collected either at the roadside or, in the case of expressways, by data collectors in vehicles. NOPUS derives its major estimates of front-seat belt use and motorcycle helmet use from the MT survey. NHTSA published the front-seat belt use results from the 2011 NOPUS MT survey in December 2011.[1] In contrast, the CI study data is collected at intersections controlled by stop signs or stoplights, where vehicle occupants are observed from the roadside. Only stopped vehicles are observed due to time constraints restricting the amount of time available to collect the variety of information required by the survey. NOPUS derives its estimates of rear-seat belt use, child restraint use, driver electronic device use, and demographic characteristics of the vehicle occupants from the CI study.

Only motorcycles and passenger vehicles without commercial or government markings are observed by the NOPUS (NOPUS does not record restraint use data for occupants of commercial vehicles, buses, taxis, or emergency vehicles). The population of interest includes all 50 States, the District of Columbia, with the sample observation sites consisting of Federal, State, county highways, residential streets, and rural roads. Data is collected only during daylight hours when light is adequate to observe seat belt use through the vehicle windshield.

The 2011 NOPUS data collection was conducted between 7 a.m. and 6 p.m. during the period from June 6, 2011, to June 17, 2011. The 2011 NOPUS survey data is based on the results of 54,475 occupants observed in the 38,215 vehicles at the 1,356 data collection sites. Of these observed occupants, 2,859 were children under 8. More details on the NOPUS sampling, data collection and estimation are discussed in Section 5: NOPUS Methodology.

The purpose of this report is to present occupant restraint use results from the 2011 National Occupant Protection Use Survey Controlled Intersection Study. NHTSA will publish results on driver electronic device use from the same survey in a separate research note. In the years prior to 2009, NHTSA usually presented the results from the NOPUS CI study through three or four Research Notes, each of which covers one specific topic. This 2011 report, like its 2010 counterpart,[2] will combine as much data as possible from the 2011 NOPUS CI study for the convenience of data users. However, in order to be consistent with the publications released by NHTSA in the years before 2009, sections in this report are arranged to cover similar topics to those in the Research Notes published previously.[3,4,5]

Please note that the terms "significant" and "statistically significant" are used interchangeably throughout this report. "Significant" always means "statistically significant" and the statistical significance level is chosen to be 0.1.

2. Demographic Results

The national seat belt use estimate was 84 percent in 2011, statistically unchanged from 85 percent in 2010.[1] This section presents the demographic breakdown of the occupants who used seat belts in 2011.

Although the NOPUS CI data is collected solely from vehicles stopped at intersections controlled by stop signs or stoplights, the estimates in this publication concerning seat belt use in the front seat reflect use by occupants in transit on all types of roadways. This is accomplished by making adjustments using data from the MT survey that observes seat belt use in vehicles in transit on general roadways.

Table 1 on page 5 presents results of passenger vehicle occupant seat belt use by demographic and other characteristics in 2010 and 2011, as well as the changes between the two years. Some major results are highlighted below.

Age

In 2011, there was no significant change in seat belt use by age group as compared to 2010: 8 to 15 years old, 16 to 24 years old, 25 to 69 years old, 70 and older. Figure 1 shows a comparison of the seat belt use rates between 2010 and 2011 among these age groups.

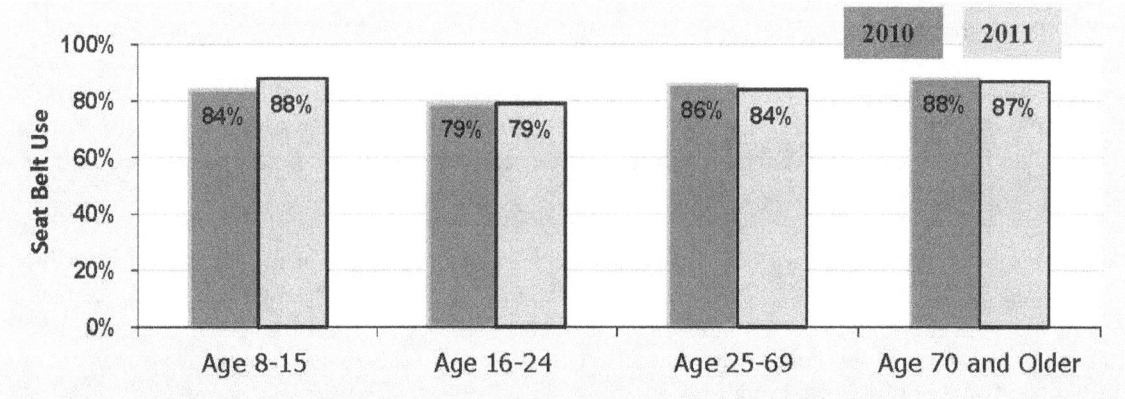

Figure 1: Seat Belt Use by Age for Occupants 8 and Older in 2010 and 2011

Figure 2 displays the trends of seat belt use for the four age groups over a period of 10 years (2002 to 2011). It shows that in 2011, seat belt use continued to be lower among 16- to 24-year-olds than other age groups.

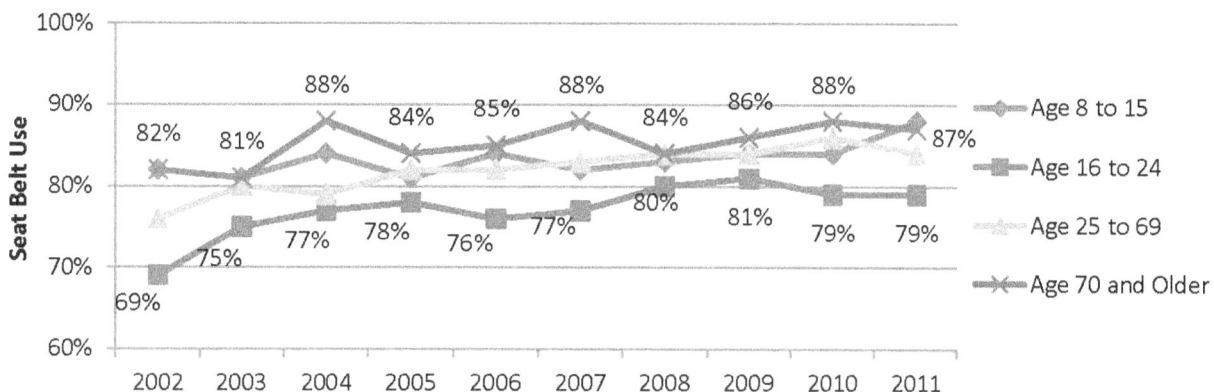

Figure 2: Seat Belt Use by Age for Occupants 8 and Older, 2002-2011

Gender

Figure 3 shows the trends of seat belt use among male and female occupants over a period of 10 years (2002 to 2011). In 2011, seat belt use continued to be lower among males (81%) than females (86%).

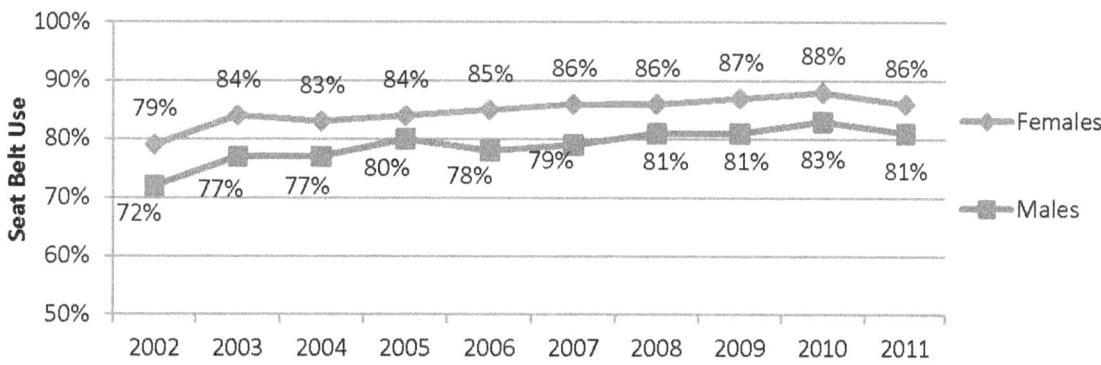

Figure 3: Seat Belt Use by Gender for Occupants 8 and Older, 2002-2011

Race

In NOPUS, vehicle occupant race is recorded as: black, white, and members of other races. The characterization is based on the visual assessment by the data collectors who observe vehicle occupants from roadsides.

Figure 4 shows the trends of seat belt use among occupants who are white, black, and members of other races over a period of 7 years (2005 to 2011). In 2011, seat belt use continued to be lower among black occupants than occupants of the other race groups. Seat belt use for members of other races was significantly higher than the complementary group (white and black occupants combined).

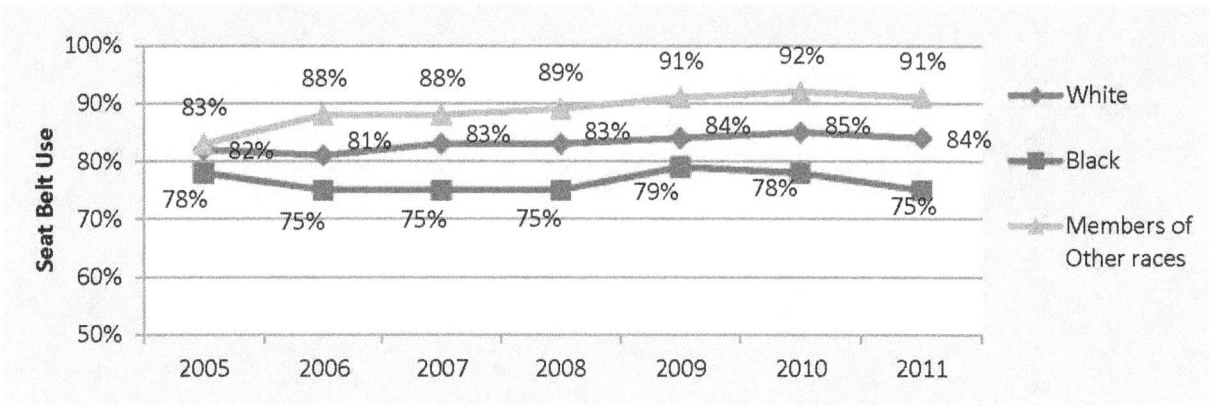

Figure 4: Seat Belt Use by Race for Occupants 8 and Older, 2005-2011

Presence of Passengers and Seat Belt Use

Figure 5 displays a clear pattern that seat belt use continued to be lower among drivers driving alone than among drivers driving with passengers.

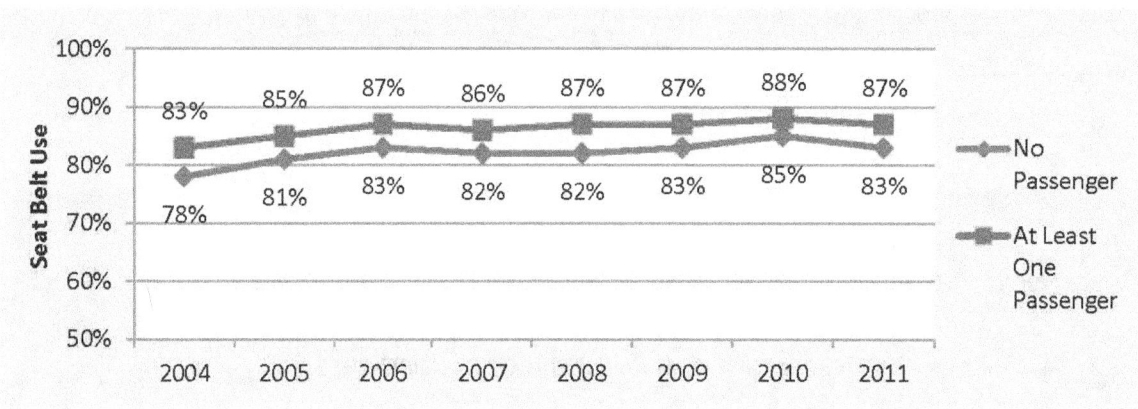

Figure 5: Passenger Effect on Seat Belt Use for Occupants 8 and Older, 2004-2011

Table 1: Passenger Vehicle Occupant Seat Belt Use by Demographic and Other Characteristics

Occupant Group[1]	2010 Belt Use[2]	2010 Confidence That Use Is High or Low in Group[3]	2011 Belt Use[2]	2011 Confidence That Use Is High or Low in Group[3]	2010 - 2011 Change in Percentage Points	2010 - 2011 Confidence in a Change in Percentage[4]
All Occupants	85%		84%		-1	77%
Males[5]	83%	**100%**	81%	**100%**	-2	67%
Females[5]	88%	**100%**	86%	**100%**	-2	86%
Occupants by Age Group[5]						
Age 8 to 15	84%	74%	88%	**98%**	4	80%
Age 16 to 24	79%	**100%**	79%	**100%**	0	14%
Age 25 to 69	86%	**100%**	84%	**94%**	-2	79%
Age 70 and Older	88%	**100%**	87%	**93%**	-1	55%
Occupants by Race[5]						
White	85%	54%	84%	61%	-1	73%
Black	78%	**100%**	75%	**100%**	-3	74%
Members of Other Races	92%	**100%**	91%	**100%**	-1	34%
Drivers With						
No Passengers	85%	**100%**	83%	**100%**	-2	83%
At Least One Passenger	88%	**100%**	87%	**100%**	-1	64%
Drivers With						
No Passengers	85%	**100%**	83%	**100%**	-2	83%
Passengers All Under Age 8	89%	**100%**	86%	72%	-3	69%
Passengers All Age 8 and Older	88%	**100%**	87%	**100%**	-1	57%
Some Passengers Under Age 8 and Some Age 8 or Older	90%	**100%**	90%	**100%**	0	7%
Drivers Age 16-24 With						
No Passengers	79%	76%	80%	83%	1	24%
Passengers All Age 16-24	77%	87%	78%	83%	1	20%
At Least One Passenger Not Age 16-24	86%	**100%**	87%	**100%**	1	13%
Occupants Age 16-24 When						
All Occupants Are Age 16-24	78%	**100%**	78%	**99%**	0	1%
At Least One Occupant Is Not Age 16-24	84%	**100%**	82%	**99%**	-2	66%

[1] Drivers and right-front passengers of passenger vehicles with no commercial or government markings.
[2] Use of shoulder belts observed between 7 a.m. and 6 p.m.
[3] The statistical confidence that use in the occupant group (e.g., occupants who are members of other races) is higher or lower than use in the corresponding complementary occupant groups (e.g., combined black and white occupants). Confidences that meet or exceed 90 percent are formatted in boldface type. Confidences are rounded to the nearest percentage point, and so confidences reported as "100 percent" are between 99.5 percent and 100 percent.
[4] The degree of statistical confidence that the 2011 use rate is different from the 2010 rate. Confidences that meet or exceed 90 percent are formatted in boldface type.
[5] The age, gender, and racial classifications are based on the subjective assessments of roadside observers.
Source: NOPUS

3. Seat Belt Use in Rear Seats

Not all vehicles on the road today have shoulder belts in the rear seats. Based on the 2010 vehicle registration data from the National Vehicle Population Profile, R.L. Polk & Co., we estimated that 92 percent of passenger vehicles on the road have shoulder belts in the rear outboard seating positions. Of the 8 percent of vehicles that have only lap belts in the rear outboard seats, all rear-seat vehicle occupants are counted by NOPUS as *not using shoulder belts*, regardless of whether they are using lap belts. Consequently, NOPUS rear-seat shoulder belt use estimates reflect both the degree to which vehicle occupants use restraints and the availability of shoulder belts in these seating positions.

Please note that rear-seat occupants might be underestimated in NOPUS because NOPUS only observes up to two passengers in the second row of seats and none in the third row and beyond.

Table 3 on page 8 presents results of seat belt use in the rear seat of passenger vehicles in 2010 and 2011 as well as the changes between the two years. Some major results are highlighted below.

Seat Belt Use in Rear Seats Versus in Front Seats

Figure 6 displays the trends of seat belt use in rear and front seats over a period of 8 years (2004 to 2011). It shows that, as in the previous years, seat belt use in 2011 was lower in the rear seat than in the front seat.

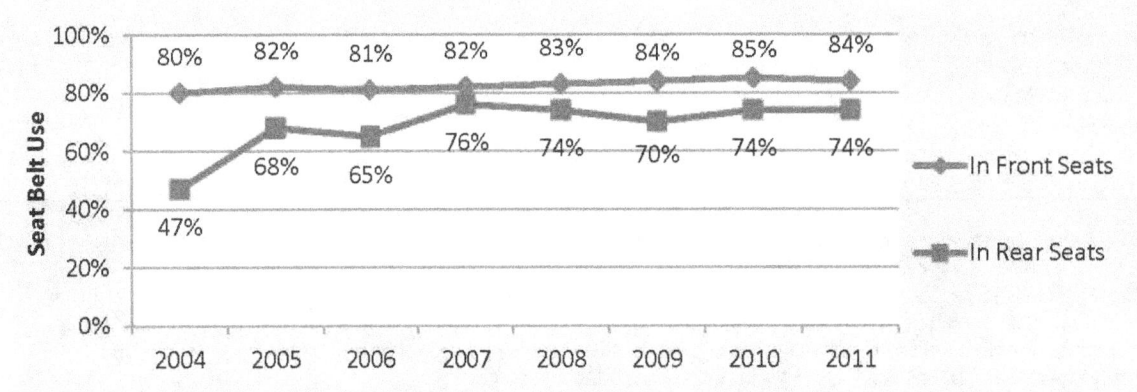

Figure 6: Seat Belt Use by Seating Position for Occupants 8 and Older, 2004-2011

State Laws and Rear-Seat Belt Use

At the time the 2011 NOPUS survey was conducted, 25 States and the District of Columbia required all vehicle occupants 18 and older to use seat belts when riding in the rear seat. Please note that rear-seat belt use laws are secondary in Kansas, New Jersey, North Carolina, Idaho, Massachusetts, Montana, Nevada, Utah, Vermont, and Wyoming. Secondary seat belt laws state that law enforcement officers may issue a ticket for not wearing a seat belt only when there is another citable traffic infraction. New Jersey's secondary law on belt use in rear seats took effect in January 2011. Table 2 provides a list of States requiring seat belts be used in all seating positions.

Table 2: States With Laws Requiring Seat Belts Be Used in All Seating Positions

Alaska	California	Delaware
District of Columbia	Idaho	Indiana
Kansas	Kentucky	Louisiana
Maine	Massachusetts	Minnesota
Montana	Nevada	New Jersey
New Mexico	North Carolina	Oregon
Rhode Island	South Carolina	Texas
Utah	Vermont	Washington
Wisconsin	Wyoming	

States with laws in effect as of June 30, 2011, requiring people 18 and older to use seat belts in all seating positions. Also includes the District of Columbia. The rear-seat seat belt use law took effect in New Jersey during the period July 1, 2010 – June 30, 2011.

Figure 7 shows the trends of rear-seat belt use among passengers in the States with or without laws requiring belt use in all seating positions over a period of 7 years (2005 to 2011). As in the previous years, seat belt use in rear seats in 2011 was higher in the States with laws requiring belt use in all seating positions (83%) than in the States requiring belt use only in the front seat (67%).

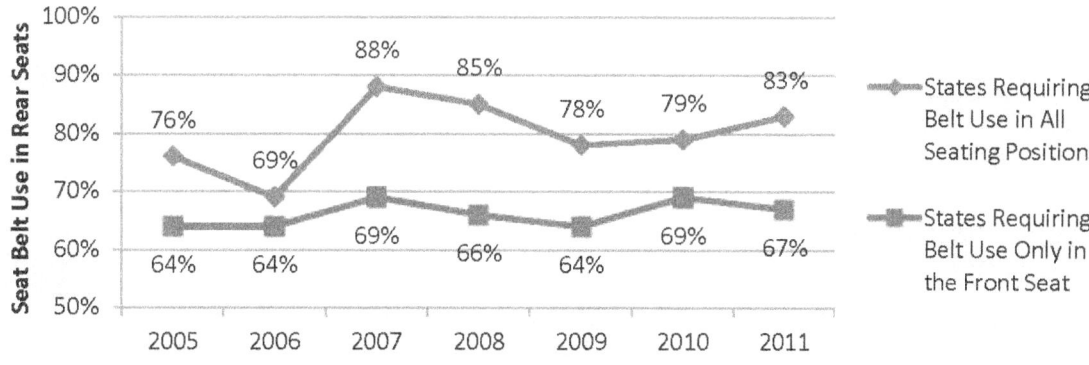

Figure 7: Seat Belt Use in Rear Seats by State Law Type for Occupants 8 and Older, 2005-2011

Table 3: Seat Belt Use in the Rear Seat of Passenger Vehicles, by Major Characteristics

Passenger Group[1]	2010		2011		2010 – 2011 Change	
	Belt Use[2]	Confidence That Use Is High or Low in Group[3]	Belt Use[2]	Confidence That Use Is High or Low in Group[3]	Change in Percentage Points	Confidence in a Change in Percentage[4]
All Passengers	74%		74%		0	22%
Males[5]	73%	78%	73%	86%	0	1%
Females[5]	74%	78%	76%	86%	2	35%
Passengers by Age Group[5]						
Age 8 to 15	75%	79%	80%	**99%**	5	74%
Age 16 to 24	73%	57%	71%	**94%**	-2	51%
Age 25 to 69	71%	**94%**	70%	**92%**	-1	14%
Age 70 and Older	81%	**97%**	73%	59%	-8	70%
Passengers by Race[5]						
White	75%	**97%**	76%	**99%**	1	34%
Black	53%	**100%**	57%	**100%**	4	42%
Members of Other Races	79%	**96%**	76%	61%	-3	54%
Passengers in States With Laws Requiring Belts Be Used						
In All Seating Positions	79%	**99%**	83%	**100%**	4	72%
In the Front Seat Only	69%	**99%**	67%	**100%**	-2	41%

[1] Up to two passengers observed in the second row of seats in passenger vehicles with no commercial or government markings.
[2] Use of shoulder belts observed between 7 a.m. and 6 p.m.
[3] The statistical confidence that use in the passenger group (e.g., passengers who are members of other races) is higher or lower than use in the corresponding complementary passenger groups (e.g., combined black and white passengers). Confidences that meet or exceed 90 percent are formatted in boldface type. Confidences are rounded to the nearest percentage point, and so confidences reported as "100 percent" are between 99.5 percent and 100.0 percent.
[4] The degree of statistical confidence that the 2011 use rate is different from the 2010 rate. Confidences that meet or exceed 90 percent are formatted in boldface type.
[5] The age, gender, and racial classifications are based on the subjective assessments of roadside observers.
Source: NOPUS

4. Child Restraint Use

In 2011, NOPUS continued to collect roadside observational data on child restraint use for all children under 8 years old. Table 5 on page 13 presents results of child restraint use in passenger motor vehicles by major characteristics in 2010 and 2011 as well as the changes between the two years. Table 7 on page 15 divides the occupants into three age groups and reports restraint use by some other characteristics among these groups. Table 6 on page 14 presents results on child rear placement by major characteristics in 2010 and 2011 as well as the changes between the two years. Some of the major results of child restraint use are discussed below.

Child Restraint Use Among All Children Age Under 8

Restraint use for children under age 8 increased significantly to 91 percent in 2011 from 89 percent in 2010. Figure 8 shows the trend of child restraint use since 2002.

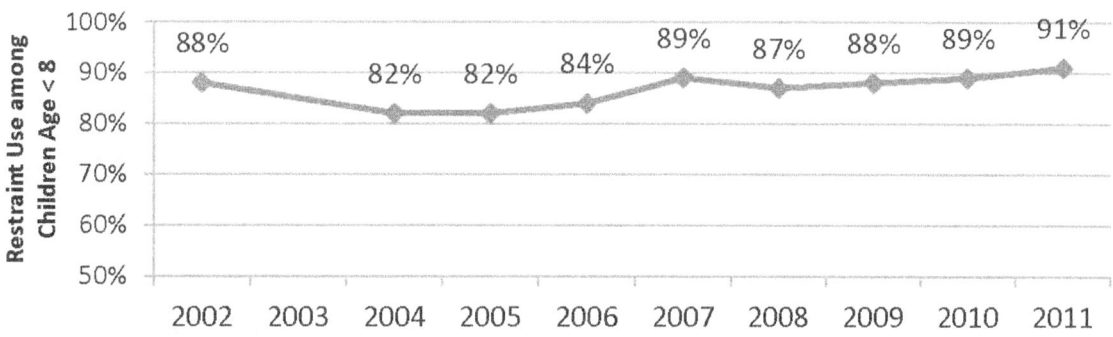

Figure 8: Child Restraint Use Among Children Under Age 8, 2002-2011

Child Rear Seat Placement

Figure 9 shows the trends of rear seat placement of children under 8 between 2002 and 2011.

The 2011 NOPUS found that 94 percent of children under 8 rode in the rear seats of vehicles. Of all the infants (from birth to 12 months), 97 percent rode in the rear seat. Ninety-nine percent of 1- to 3-year-old and 92 percent of 4- to 7-year-old children were in the rear seats in 2011. The rear seat placement rate for children age 4 to 7 increased significantly from 89 percent in 2010 to 92 percent in 2011.

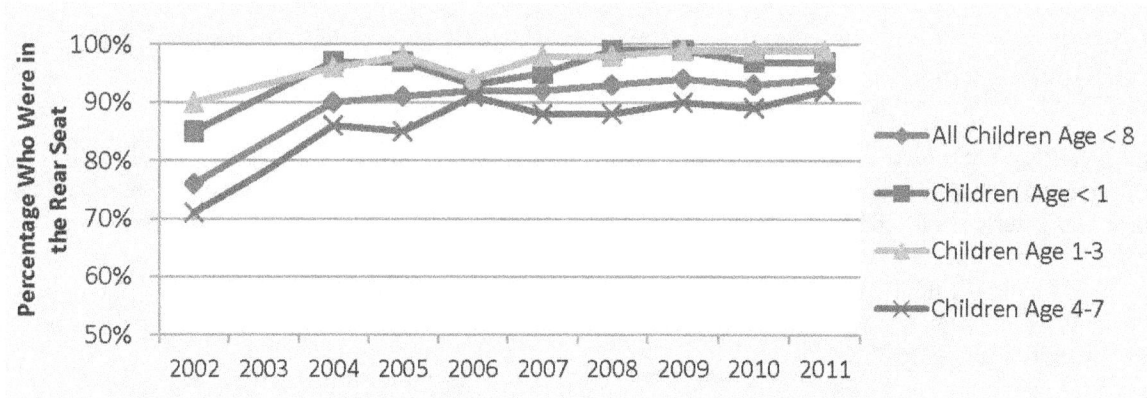

Figure 9: Child Rear Seat Placement, 2002 - 2011

At the time the 2011 survey was conducted, 9 States required children 5 and younger who weighed less than 80 pounds and were less than 54 inches tall to ride in the rear seats of vehicles. Table 4 lists the States with child rear placement laws.

Table 4: States With Laws Requiring Children 5 and Younger Be in the Rear Seat*

California	Georgia	Maine
New Jersey	Rhode Island	South Carolina
Tennessee	Washington	Wyoming

* Among children less than 80 pounds and less than 54" tall. States with laws in effect as of June 30, 2011. In no other States did such laws take effect during the period July 1, 2010, to June 30, 2011. In Delaware, children 11 and younger and 65 inches or less must be the rear seat if passenger air bag is active.

Child Restraint Use by Region

Child restraint use in the Northeast increased significantly from 89 percent in 2010 to 94 percent in 2011. Figure 10 shows the significant increase in the Northeast.

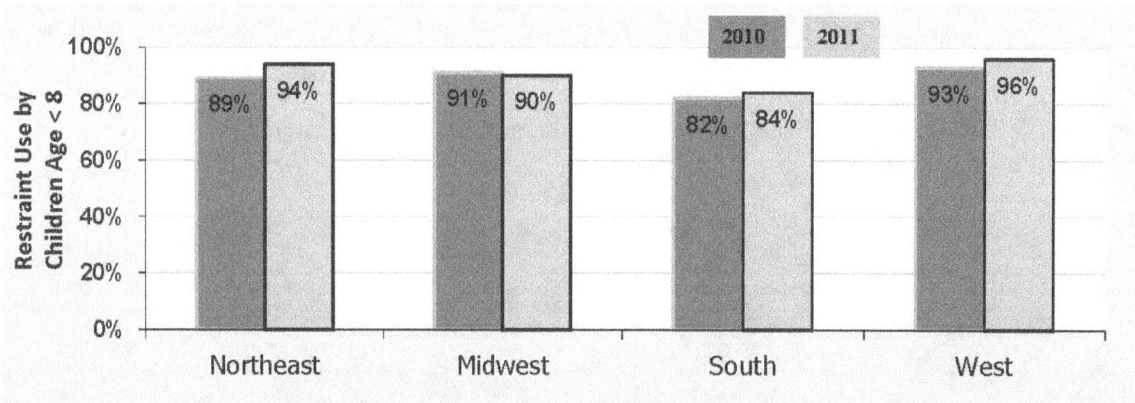

Figure 10: Child Restraint Use by Region in 2010 and 2011

As shown in Figure 11, child restraint use continued to be higher in the West than in the other regions in 2011.

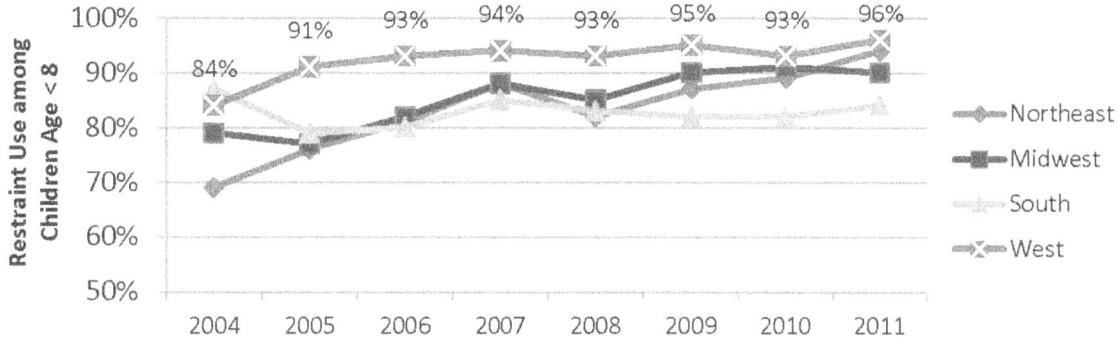

Figure 11: Child Restraint Use by Region, 2004-2011

Child Restraint Use by Time of Week

As shown in Figure 12, restraint use for children traveling during weekends increased significantly from 87 percent in 2010 to 94 percent in 2011.

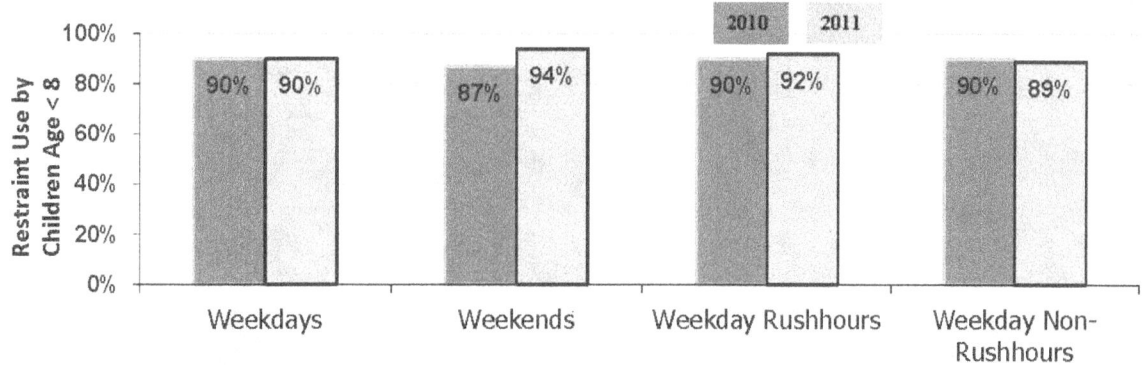

Figure 12: Child Restraint Use by Time of Week in 2010 and 2011

Child Restraint Use by Vehicle Type

As shown in Figure 13, restraint use for children traveling in vans and SUVs increased significantly from 92 percent in 2010 to 97 percent in 2011.

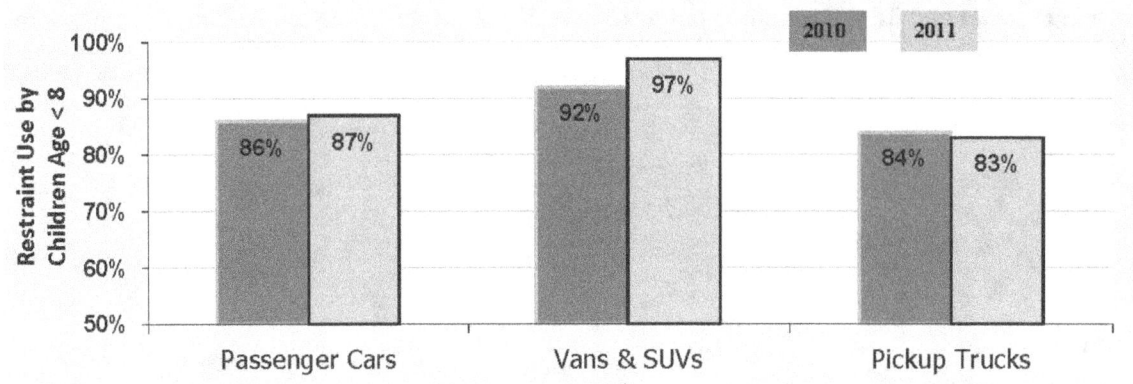

Figure 13: Child Restraint Use by Vehicle Type in 2010 and 2011

Child Restraint Use by Driver Belt Status

As shown in Figure 14, restraint use for children driven by belted drivers continued to be statistically significantly higher than for those driven by unbelted drivers.

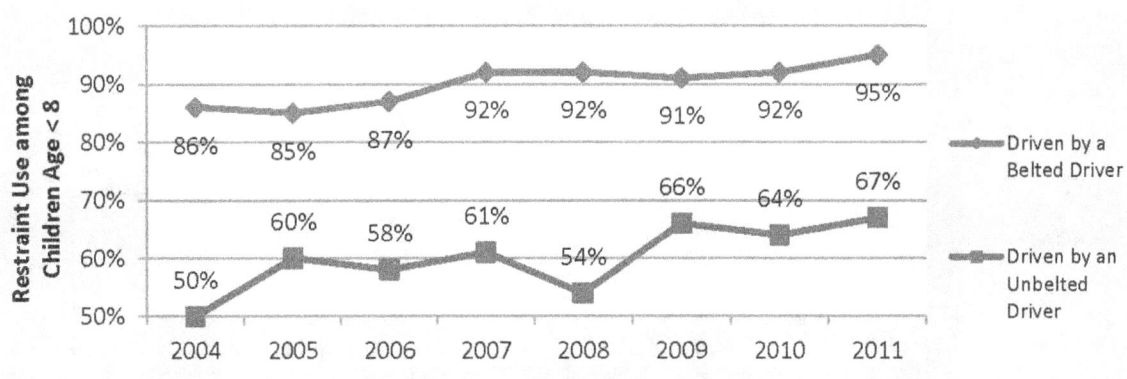

Figure 14: Child Restraint Use by Driver Belt Status, 2004-2011

Table 5: Child Restraint Use in Passenger Motor Vehicles, by Major Characteristics

Child Passenger Group[1]	2010		2011		2010-2011 Change	
	Restraint Use[2]	Confidence That Use Is High or Low in Group[3]	Restraint Use[2]	Confidence That Use Is High or Low in Group[3]	Change in Percentage Points	Confidence in a Change in Use[4]
All Child Passengers (From Birth to 7 Years)	89%		91%		2	**94%**
Children Driven by						
a Belted Driver	92%	**100%**	95%	**100%**	3	**100%**
an Unbelted Driver	64%	**100%**	67%	**100%**	3	38%
a Male Driver	87%	**99%**	92%	82%	5	**99%**
a Female Driver	90%	**99%**	91%	82%	1	20%
a Driver Age 16 to 24	90%	65%	89%	77%	-1	11%
a Driver Age 25 to 69	89%	55%	92%	73%	3	**93%**
a Driver Age 70 and Older	85%	70%	92%	58%	7	58%
a White Driver	92%	**100%**	93%	**99%**	1	71%
a Black Driver	76%	**100%**	75%	**100%**	-1	11%
a Driver who is a Member of Other Races	84%	**96%**	93%	72%	9	**97%**
Children in						
the Front Seat	81%	**99%**	72%	**100%**	-9	82%
the Rear Seat	90%	**99%**	93%	**100%**	3	**98%**
Child Passengers on						
Expressways	89%	64%	95%	**99%**	6	**100%**
Surface Streets	89%	64%	89%	**99%**	0	20%
Child Passengers Traveling in						
Fast Traffic	90%	78%	93%	89%	3	**93%**
Medium-Speed Traffic	88%	85%	90%	73%	2	71%
Slow Traffic	90%	71%	89%	72%	-1	9%
Child Passengers in						
Passenger Cars	86%	**100%**	87%	**100%**	1	25%
Vans & SUVs	92%	**100%**	97%	**100%**	5	**100%**
Pickup Trucks	84%	**93%**	83%	89%	-1	15%
Child Passengers in the						
Northeast	89%	51%	94%	**93%**	5	**97%**
Midwest	91%	85%	90%	62%	-1	10%
South	82%	**99%**	84%	**99%**	2	32%
West	93%	**100%**	96%	**100%**	3	80%
Child Passengers in						
Urban Areas	80%	**100%**	92%	53%	12	**99%**
Suburban Areas	91%	**100%**	92%	64%	1	24%
Rural Areas	89%	51%	90%	67%	1	33%
Child Passengers Traveling During						
Weekdays	90%	**94%**	90%	**100%**	0	18%
Rush Hours	90%	55%	92%	86%	2	60%
Nonrush Hours	90%	55%	89%	86%	-1	55%
Weekends	87%	**94%**	94%	**100%**	7	**100%**

[1] Passengers under age 8 observed between 7 a.m. and 6 p.m. in the right-front seat or the second row of seats in passenger vehicles with no commercial or government markings that are stopped at a stop sign or stop light. Age, gender, and racial classifications are based on the subjective assessments of roadside observers.
[2] Use of child car seats (forward- or rear-facing), booster seats, and seat belts.
[3] The statistical confidence that use in the passenger group (e.g., child passengers in the Northeast) is higher or lower than use in the corresponding complementary passenger group (e.g., combined child passengers in the Midwest, in the South and in the West). Confidences that meet or exceed 90 percent are formatted in boldface type. Confidences are rounded to the nearest percentage point, and so confidences reported as "100 percent" are between 99.5 percent and 100.0 percent.
[4] The degree of statistical confidence that the 2011 use rate is different from the 2010 rate. Confidences that meet or exceed 90 percent are formatted in boldface type.
Source: NOPUS

Table 6: The Percent of Children Who Rode in the Rear Seat, by Major Characteristics

Child Passenger Group[1]	2010		2011		2010-2011 Change	
	Percentage Who Were in Rear Seat[2]	Confidence That Use Is High or Low in Group[3]	Percentage Who Were in Rear Seat[2]	Confidence That Use Is High or Low in Group[3]	Change in Percentage Points	Confidence in a Change in Rear Seat Occupancy[4]
All Child Passengers (From Birth to 7 Years)	93%		94%		1	77%
Age 0 (Infants)	97%	96%	97%	99%	0	5%
Age 1-3	99%	100%	99%	100%	0	2%
Age 4-7	89%	100%	92%	100%	3	**91%**
Child Passengers in States With[5]						
Law Requiring Children From Birth of 5 Years Be in the Rear Seat	94%	89%	95%	72%	1	31%
No Such Law	92%	89%	94%	72%	2	79%
Children Driven by						
a Belted Driver	93%	**93%**	95%	**98%**	2	79%
an Unbelted Driver	88%	**93%**	90%	**98%**	2	34%
a Male Driver	93%	75%	96%	**93%**	3	**87%**
a Female Driver	93%	75%	93%	**93%**	0	43%
a Driver Age 16 to 24	98%	**100%**	96%	85%	-2	69%
a Driver Age 25 to 69	92%	**100%**	94%	55%	2	**92%**
a Driver Age 70 and Older	94%	62%	89%	79%	-5	48%
a White Driver	93%	80%	94%	89%	1	63%
a Black Driver	91%	77%	95%	60%	4	60%
a Driver who is a Member of Other Races	95%	**99%**	96%	**91%**	1	61%
Child Passengers on						
Expressways	95%	**100%**	97%	**100%**	2	62%
Surface Streets	92%	**100%**	93%	**100%**	1	51%
Child Passengers Traveling in						
Fast Traffic	95%	**99%**	94%	51%	-1	19%
Medium-Speed Traffic	92%	**95%**	94%	60%	2	**90%**
Slow Traffic	92%	66%	95%	57%	3	65%
Child Passengers in						
Passenger Cars	94%	86%	94%	59%	0	13%
Vans & SUVs	93%	73%	96%	**99%**	3	**92%**
Pickup Trucks	82%	**100%**	82%	**99%**	0	8%
Child Passengers in the						
Northeast	96%	**99%**	97%	**96%**	1	33%
Midwest	92%	75%	93%	79%	1	32%
South	90%	**96%**	91%	**99%**	1	41%
West	94%	83%	96%	**94%**	2	76%
Child Passengers in						
Urban Areas	94%	72%	95%	78%	1	27%
Suburban Areas	93%	65%	96%	**98%**	3	**96%**
Rural Areas	92%	88%	91%	**97%**	-1	31%
Child Passengers Traveling During						
Weekdays	93%	61%	93%	**100%**	0	9%
Rush Hours	92%	**91%**	93%	53%	1	53%
Nonrush Hours	94%	**91%**	93%	53%	-1	52%
Weekends	93%	61%	97%	**100%**	4	**96%**
Child Passengers in a						
Rear-Facing Car Seat	100%	**100%**	99%	**100%**	-1	51%
Forward-Facing Car Seat	99%	**100%**	99%	**100%**	0	3%
High-Backed Booster Seat	100%	**100%**	98%	**100%**	-2	82%
Seat belt or Backless Booster Seat	86%	**100%**	92%	**100%**	6	**99%**
No Restraint Observed	88%	**98%**	82%	**100%**	-6	75%

[1] Passengers under 8 observed between 7 a.m. and 6 p.m. in the right-front seat or the second row of seats in passenger vehicles with no commercial or government markings that are stopped at a stop sign or stoplight. Age, gender, and racial classifications are based on the subjective assessments of roadside observers.

[2] The percentage of the child passenger group who were in the second row of seats at the time of observation.

[3] The statistical confidence that use in the passenger group (e.g., child passengers in the Northeast) is higher or lower than use in the corresponding complementary passenger group (e.g., combined child passengers in the Midwest, in the South and in the West). Confidences that meet or exceed 90 percent are formatted in boldface type. Confidences are rounded to the nearest percentage point, and so confidences reported as "100 percent" are between 99.5 percent and 100.0 percent.

[4] The degree of statistical confidence that the percentage of the child passenger group who were in the rear seat in 2011 is different from the analogous percentage from 2010.

[5] Use rates reflect the law in effect at the time data was collected.

Source: NOPUS

Table 7: Child Restraint Use in Passenger Motor Vehicles, by Age and Other Characteristics

Child Passenger Group[1]	2010		2011		2010-2011 Change	
	Restraint Use[2]	Confidence That Use Is High or Low in Group[3]	Restraint Use[2]	Confidence That Use Is High or Low in Group[3]	Change in Percentage Points	Confidence in a Change in Use[4]
Infants (From Birth to 12 Months)						
Infants Driven by						
a Belted Driver	99%	84%	99%	87%	0	1%
an Unbelted Driver	94%	84%	97%	87%	3	34%
a Male Driver	97%	**98%**	99%	70%	2	55%
a Female Driver	100%	**98%**	99%	70%	-1	79%
Infants in						
Passenger Cars	99%	86%	98%	**98%**	-1	83%
Vans & SUVs	99%	56%	100%	**98%**	1	82%
Pickup Trucks	NA	NA	NA	NA	NA	NA
Infants in the						
Northeast	97%	87%	100%	87%	3	78%
Midwest	100%	**99%**	96%	**96%**	-4	**96%**
South	99%	68%	99%	61%	0	25%
West	99%	69%	100%	**98%**	1	68%
Infants in						
Urban Areas	99%	63%	99%	78%	0	42%
Suburban Areas	99%	61%	99%	84%	0	60%
Rural Areas	99%	57%	98%	86%	-1	51%
Children Age 1 to 3						
Children Age 1-3 Driven by						
a Belted Driver	95%	**99%**	97%	**98%**	2	**95%**
an Unbelted Driver	80%	**99%**	85%	**98%**	5	38%
a Male Driver	93%	89%	96%	54%	3	77%
a Female Driver	95%	89%	96%	54%	1	34%
Children Age 1-3 in						
Passenger Cars	92%	**99%**	94%	**95%**	2	53%
Vans & SUVs	96%	**99%**	97%	**96%**	1	64%
Pickup Trucks	94%	56%	97%	66%	3	44%
Children Age 1-3 in the						
Northeast	94%	57%	96%	51%	2	32%
Midwest	96%	**93%**	96%	62%	0	3%
South	89%	**93%**	93%	**91%**	4	47%
West	96%	**93%**	97%	88%	1	31%
Children Age 1-3 in						
Urban Areas	86%	**95%**	94%	71%	8	79%
Suburban Areas	96%	**93%**	95%	65%	-1	20%
Rural Areas	95%	65%	97%	88%	2	74%
Children Age 4 to 7						
Children Age 4-7 Driven by						
a Belted Driver	87%	**100%**	92%	**100%**	5	**99%**
an Unbelted Driver	51%	**100%**	51%	**100%**	0	7%
a Male Driver	82%	89%	89%	**91%**	7	**99%**
a Female Driver	85%	89%	86%	**91%**	1	35%
Children Age 4-7 in						
Passenger Cars	79%	**100%**	81%	**100%**	2	31%
Vans & SUVs	88%	**100%**	96%	**100%**	8	**100%**
Pickup Trucks	79%	83%	73%	**91%**	-6	38%
Children Age 4-7 in the						
Northeast	83%	52%	92%	**96%**	9	**99%**
Midwest	87%	**92%**	86%	58%	-1	7%
South	74%	**100%**	77%	**97%**	3	28%
West	89%	**99%**	93%	**99%**	4	84%
Children Age 4-7 in						
Urban Areas	72%	**99%**	88%	59%	16	**99%**
Suburban Areas	86%	**96%**	88%	58%	2	43%
Rural Areas	84%	66%	86%	66%	2	28%

[1] Passengers under 8 observed between 7 a.m. and 6 p.m. in the right-front seat or the second row of seats in passenger vehicles with no commercial or government markings that are stopped at a stop sign or stoplight. Age, gender, and racial classifications are based on the subjective assessments of roadside observers.
[2] Use of child car seats (forward- or rear-facing), booster seats, and seat belts.
[3] The statistical confidences that use in the passenger group (e.g., child passengers in the Northeast) is higher or lower than use in the corresponding complementary passenger group (e.g., combined child passengers in the Midwest, in the South and in the West). Confidences that meet or exceed 90 percent are formatted in boldface type. Confidences are rounded to the nearest percentage point, and so confidences reported as "100 percent" are between 99.5 percent and 100.0 percent.
[4] The degree of statistical confidence that the 2011 use rate is different from the 2010 rate. Confidences that meet or exceed 90 percent are formatted in boldface type.
NA: Data not sufficient to produce a reliable estimate.
Source: NOPUS

5. NOPUS Methodology

This section briefly discusses the sample design, data collection, and estimation used in the 2011 NOPUS Controlled Intersection Study. Data collection, estimation, and variance estimation for NOPUS are conducted by Westat, Inc., under the direction of NHTSA's National Center for Statistics and Analysis under Federal contract number DTNH22-07-D-00057.

Sample Design

The NOPUS uses a complex multistage probability sample, statistical data editing, imputation of unknown values, and complex estimation procedures. The sample sites for the 2011 NOPUS were entirely from the 2006 NOPUS sample redesign without incorporating any sites from the old design. During the transitional years between 2006 and 2010, sample sites were chosen both from the new design and the old design. Prior to 2006, sample sites were from the old design only.

The NOPUS sample was selected using a two-stage design with stratified probability proportional to size (PPS) sampling at each stage. The sampling frame of PSUs for the 2006 redesigned sample included all counties in the U.S. but excluded Puerto Rico and the U.S. Territories. In the redesigned sample, only one PSU was designated as a certainty sampling unit (i.e., probability one) due to its large vehicle miles traveled (VMT). In order to decrease the variances associated with the survey estimates, the remaining PSUs were stratified according to their predicted rates of restraint use based on a regression model that used primary enforcement law status, ratio of fatal crashes to VMT, percentage of college graduates, and several other relevant variables as predictors. The non-certainty PSUs were selected by systematic PPS sampling from these primary strata using VMT as the measure of size. The secondary sampling units (SSUs) consisted of road segments that lie at least partly inside the selected PSUs. To define road segments, the selected PSUs were divided into grids, usually of one-acre in size.

Table 8 shows the observed sample sizes of the 2011 NOPUS. A total of 54,475 occupants were observed in the 38,215 vehicles at the 1,356 data collection sites. Of these observed occupants, 2,859 were children under 8. Please note that due to ineligibility, construction, danger in the area, or road closure, observations could not be completed at some of the sampled observation sites.

Table 8: Sites, Vehicles and Occupants in the 2011 NOPUS

Numbers of	2010	2011	Percentage Change
Sites Observed	1,446	1,356	-6%
Vehicles Observed	48,331	38,215	-21%
Occupants Age 8 and Older	65,833	51,616	-22%
In Front Seat	62,349	48,890	-22%
In Rear Seat	3,484	2,726	-22%
Occupants Under Age 8	3,914	2,859	-27%
Children Under Age 1	540	383	-29%
Children Age 1 to 3	1,317	937	-29%
Children Age 4 to 7	2,057	1,539	-25%

Data Collection

The 2011 NOPUS data collection was conducted during the period from June 6, 2011, to June 17, 2011.

In the NOPUS Controlled Intersection Study, trained data collectors observe restraint use of drivers and other occupants of passenger vehicles having no commercial or government markings which have stopped at a stop sign or stoplight during daylight hours between 7 a.m. and 6 p.m. Observations are made both on the surface streets and at the ends of the expressway exit ramps (where there are controlled intersections.) Only stopped vehicles are observed based on the time required to collect the variety of information required by the survey, including subjective assessments of the vehicle occupants' age and race. Observers collect data on the driver, right-front passenger, and up to two passengers in the second row of seats. Observers do not interview vehicle occupants intentionally, allowing NOPUS to capture the uninfluenced behavior of the occupants.

NOPUS Controlled Intersection Study is always done following NOPUS Moving Traffic Survey and is usually scheduled for all surface streets and limited access highway ramps, where NOPUS data from previous years indicates that a controlled intersection exists. If the data collectors arrive at an assigned surface street site and the site is not controlled, they are instructed to search for an alternative. The data collectors move down the roadside and record vehicle and occupant characteristics. Once the traffic light turns green or they finish observing all vehicles, the data collectors return to the intersection to wait for the next traffic light cycle or next vehicle. They observe vehicles in the lane closest to their observational position, even if the closest lane is an exclusive turn lane (which is often the case at the controlled intersections.) When possible and if visibility allows, the data collectors also observe the other lanes of traffic. The data collectors are instructed to record the first behavior of the driver in which they observe.

Regardless of road type, the data collectors observe vehicles at the assigned intersections for 40 minutes. Since data collection for the CI study immediately follows the MT survey, no additional vehicle counts are conducted at controlled intersections. Instead, the independent counts from the MT survey observation sites are used for the corresponding CI study sites.

Estimation

NOPUS estimates the rate of occupants restrained in restraint type (R) among the occupants having characteristic (C) using the formula,

$$\text{Restraint Use}_{CR} = \frac{\sum_{i,j,k} w_{ijk} F_{ijk} CR_{ijk}}{\sum_{i,j,k} w_{ijk} F_{ijk} C_{ijk}},$$

where w_{ijk} and F_{ijk}, respectively, denote the base weight and the product of various weight adjustment factors at the site k in the stratum j of the PSU i. CR_{ijk} stands for the number of observed occupants having characteristic C and restrained in restraint type R and C_{ijk} denotes the number of observed occupants having characteristic C at the site k in the stratum j of the PSU i. For example, the seat belt use by vehicle type is estimated using the above formula, where CR_{ijk} is the number of observed belted occupants in certain type of vehicles (such as passenger cars, vans and SUVs, or pickup trucks) and C_{ijk} is the number of ALL (belted and unbelted) occupants observed in that type of vehicles at the site k in the stratum j of the PSU i.

In certain instances, NHTSA does not provide estimates. These are typically restraint use estimates whose numerator is based on fewer than five persons observed, whose denominator is based on fewer than 30 people observed, or the estimates are not statistically different from 0 percent (i.e., the standard error is at least half the point estimate). These are reported as "NA" in publications. Any related estimate (i.e., change in use and confidence estimates) is not reported as well. The same criteria are used in reporting estimates from the National Survey of Use of Booster Seats (NSUBS).

6. References

[1] Pickrell, T. M., & Ye, T. J. (2011, December). *Seat Belt Use in 2011 – Overall Results*, (Report No. DOT HS 811 544). Washington, DC: National Highway Traffic Safety Administration. Available at www-nrd.nhtsa.dot.gov/Pubs/811544

[2] Pickrell, T.M., & Ye, T.J. (2012, January). *Occupant Restraint Use in 2010 – Results from the National Occupant Protection Use Survey Controlled Intersection Study.* (Report No. DOT HS 811 527). Washington, DC: National Highway Traffic Safety Administration. Available at www-nrd.nhtsa.dot.gov/Pubs/811527

[3] Pickrell, T. M., & Ye, T. J. (2009, August). *Seat Belt Use in 2008 – Demographic Results.* (Report No. DOT HS 811 183). Washington, DC: National Highway Traffic Safety Administration. Available at www-nrd.nhtsa.dot.gov/Pubs/811183

[4] Pickrell, T. M., & Ye, T. J. (2009, May). *Seat Belt Use in Rear Seats in 2008.* (Report No. DOT HS 811 133). Washington, DC: National Highway Traffic Safety Administration. Available at www-nrd.nhtsa.dot.gov/Pubs/811133

[5] Pickrell, T. M., & Ye, T. J. (2009, May). *Child Restraint Use in 2008 – Overall Results.* (Report No. DOT HS 811 135). Washington, DC: National Highway Traffic Safety Administration. Available at www-nrd.nhtsa.dot.gov/Pubs/811135

Appendix: Definitions

- Vehicle occupants observed in the NOPUS survey are counted as "belted" if they appeared to have a shoulder belt across the front of the body. NOPUS does not observe the use of lap belts because these restraints cannot be reliably observed from the roadside.

- The survey classifies a child as:
    - <u>Restrained in a rear-facing car seat</u> if the child appears to be on a seat on top of the vehicle seat, facing the rear of the vehicle, with harness straps across the front of the child.
    - <u>Restrained in a forward-facing car seat</u> if the child appears to be on a seat on top of the vehicle seat, facing the front of the vehicle, with harness straps across the front of the child.
    - <u>Restrained in a high-backed booster seat</u> if the child appears to be on a seat on top of the vehicle seat with a shoulder belt across the front of the child.
    - <u>Restrained in a seat belt or backless booster seat</u> if there is a shoulder belt across the front of the child but the observers cannot see if the child is in a seat on top of the vehicle seat.
    - <u>Restrained</u> if s/he is restrained by any of the above.
    - The remaining children are classified as <u>unrestrained</u>. Note that in the survey there is no mention of being "unrestrained" in, for example, a forward-facing car seat. NOPUS does not observe the use of lap belts, and does not distinguish between seat belts and backless booster seats, because these assessments cannot be reliable if observed from the roadside.

- The racial categories "Black," "White," and "Members of other races" in NOPUS reflect subjective characterizations by roadside observers regarding the race of vehicle occupants. Likewise observers record all age groups (8 to 15 years old, 16 to 24 years old, 25 to 69 years old, and 70 and older) that best fits their visual assessment of each observed occupant.

- "Expressways" are defined as roadways with limited access, while "surface streets" comprise all other roadways.

- A roadway is defined to have "fast traffic" if, during the observation period, the average speed of passenger vehicles passing the observer(s) exceeds 50 mph, with "medium-speed traffic" defined as 31 to 50 mph and "slow traffic" defined as 30 mph or slower. The traffic speed data in the CI survey are matched to the MT survey data.

- A roadway is defined to have "heavy traffic" if the average number of vehicles on the roadway during the observation period is greater than 5 per lane per mile, with "moderately dense traffic" defined as greater than 1 but less than or equal to 5 vehicles per lane per mile, and "light traffic" as less than or equal to 1 vehicle per lane per mile. Please note that this traffic density breakdown has been revised in the 2011 NOPUS to better capture the traffic patterns. The traffic density data in the CI survey is matched to the MT survey data.

- Since NOPUS is not a census but based on some probability sample, it is impossible to produce State-by-State restraint use results. However NOPUS can and does produce regional estimates using the following categories:

Northeast: Connecticut, Massachusetts, Maine, New Hampshire, New Jersey, New York, Pennsylvania, Rhode Island, Vermont

Midwest: Iowa, Kansas, Illinois, Indiana, Michigan, Minnesota, Missouri, North Dakota, Nebraska, Ohio, South Dakota, Wisconsin

South : Alabama, Arkansas, the District of Columbia, Delaware, Florida, Georgia, Kentucky, Louisiana, Maryland, Mississippi, North Carolina, Oklahoma, South Carolina, Tennessee, Texas, Virginia, West Virginia

West: Alaska, Arizona, California, Colorado, Hawaii, Idaho, Montana, New Mexico, Nevada, Oregon, Washington, Washington, Wyoming

These definitions of the four NOPUS regions are the same regional definitions used in the NSUBS.

DOT HS 811 697
January 2013

U.S. Department
of Transportation
**National Highway
Traffic Safety
Administration**

9205-011713-v3